THE ENVIRONMENT

THE REASON WHY SERIES

THE ENVIRONMENT

IRVING ADLER
illustrated by Peggy Adler

The John Day Company • New York

THE REASON WHY SERIES

AIR—*Revised Edition, 1972*
ATOMIC ENERGY
ATOMS AND MOLECULES
THE CALENDAR
COAL—*Revised Edition, 1974*
COMMUNICATION
DIRECTIONS AND ANGLES
THE EARTH'S CRUST
ENERGY
EVOLUTION
FIBERS—*Updated Edition, 1972*
HEAT AND ITS USES
   Revised Edition of HEAT, 1973
HOUSES
INSECTS AND PLANTS
INTEGERS: POSITIVE AND NEGATIVE
IRRIGATION: CHANGING DESERT INTO GARDEN
LANGUAGE AND MAN
LEARNING ABOUT STEEL: THROUGH THE STORY OF A NAIL
MACHINES
MAGNETS
NUMBERS OLD AND NEW
NUMERALS: NEW DRESSES FOR OLD NUMBERS
OCEANS
RIVERS
SETS
SHADOWS—*Revised Edition, 1968*
STORMS
TASTE, TOUCH AND SMELL
THINGS THAT SPIN: FROM TOPS TO ATOMS
TREE PRODUCTS
WHY? A BOOK OF REASONS
WHY AND HOW? A SECOND BOOK OF REASONS
YOUR EARS
YOUR EYES

Copyright © 1976 by Irving Adler
All rights reserved. Except for use in a review, the reproduction or utilization of this work in any form or by any electronic, mechanical, or other means, now known or hereafter invented, including xerography, photocopying, and recording, and in any information storage and retrieval system is forbidden without the written permission of the publisher. Published simultaneously in Canada by Fitzhenry & Whiteside Limited, Toronto.
Manufactured in the United States of America

Library of Congress Cataloging in Publication Data
Adler, Irving. The environment. (The Reason why series)
SUMMARY: Discusses the importance of the environment, how it is abused, and what can be done to protect it.
1. Ecology—Juvenile literature. 2. Environmental protection—Juvenile literature.
[1. Ecology. 2. Environmental protection]   I. Adler, Peggy. II. Title.
QH541.14.A34   1976   574.5   75-35526
ISBN 0-381-99617-4 RB

10 9 8 7 6 5 4 3 2 1

Contents

Our Environment—What It Is   6
Living Things in the Environment   8
Cycles in Nature   10
Food Chains   14
Connections and Balance in Nature   16
Man's Work   18
Machines, Power and Chemistry   20
Population and Its Limits   22
Unwanted Side Effects   24
Vanishing Species   26
Used-up Treasures   28
Hogging the World's Resources   30
Fouling the Air   32
Spoiled Water   34
Destruction of Soil   36
Mountains of Rubbish   38
Man-made Poisons   40
Other Environmental Problems   42
Protecting the Environment   44
What Everyone Can Do   46
Word List   47
Index   48

Our Environment—What It Is

Our environment is what surrounds us. From it we get the things we need to live and work and play.

Air is part of our environment. In it is oxygen we breathe, and nitrogen that becomes food for the plants that are food for us.

Water is part of our environment. We drink it, cook with it and bathe in it. In seas, lakes and rivers it is the home of fish we catch to eat. In the ground it helps to make plants grow.

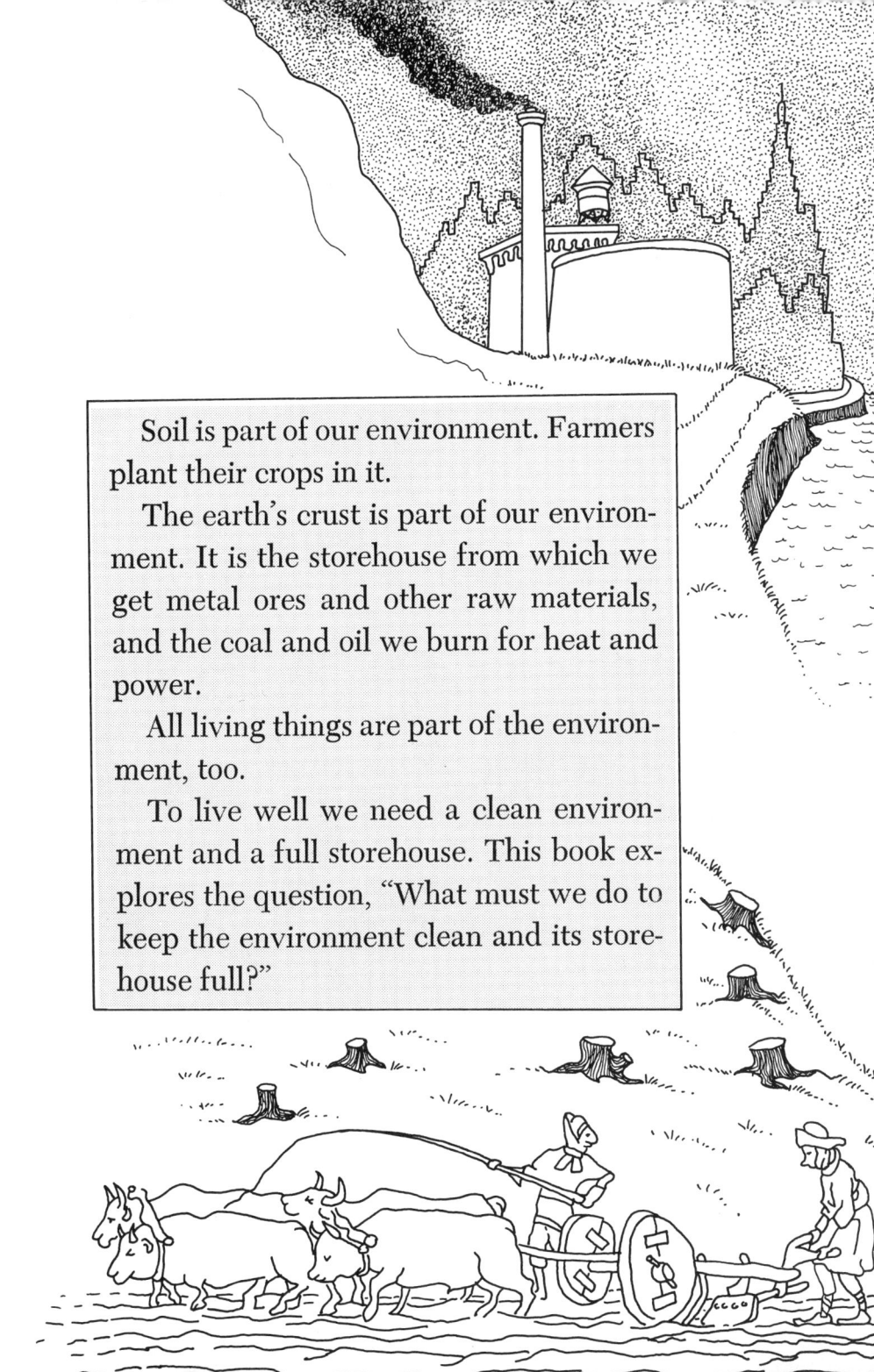

Soil is part of our environment. Farmers plant their crops in it.

The earth's crust is part of our environment. It is the storehouse from which we get metal ores and other raw materials, and the coal and oil we burn for heat and power.

All living things are part of the environment, too.

To live well we need a clean environment and a full storehouse. This book explores the question, "What must we do to keep the environment clean and its storehouse full?"

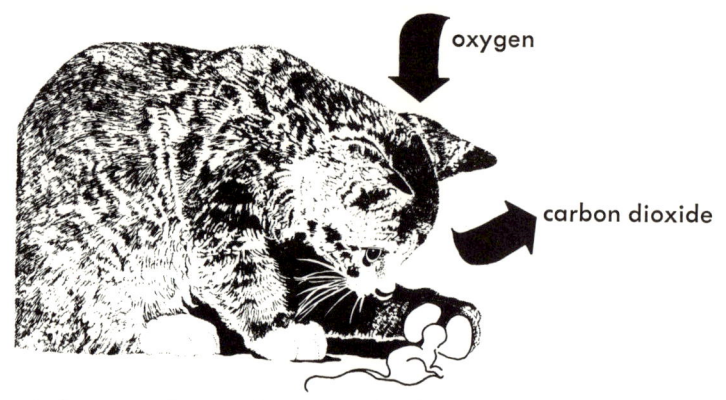

A cat changes the environment by breathing in oxygen, breathing out carbon dioxide and eating mice.

Living Things in the Environment

Living things in the environment include plants and animals. While they are part of the environment, they constantly interfere with it. Every living thing, including man, takes some things out of the environment and puts other things into it.

Every animal, for example, grows and moves. Its body is made of many chemicals, some of which have large molecules containing carbon. It needs more and more of these large molecules for two reasons. First, it uses some as building blocks, building them into its body as it grows. Second, it uses some as fuel in a kind of burning, called *respiration,* that releases the energy needed to grow and move. To get the large molecules it needs, every animal must take from the environment other animals or plants that it eats. To burn molecules used as fuel, it must also take oxygen from the environment. Respiration produces carbon dioxide and water as waste products. So all living animals take oxygen out of the environment and put carbon dioxide and water into it.

There is another form of interference with the environment that works in the opposite direction. Green plants do not get the large molecules they need by eating other plants or animals. They make them out of sugar molecules they make themselves in a process called *photosynthesis,* using energy obtained from sunlight to join molecules of carbon dioxide and water. A waste product of this process is oxygen. So a plant that carries out photosynthesis takes carbon dioxide and water out of the environment and puts oxygen into it.

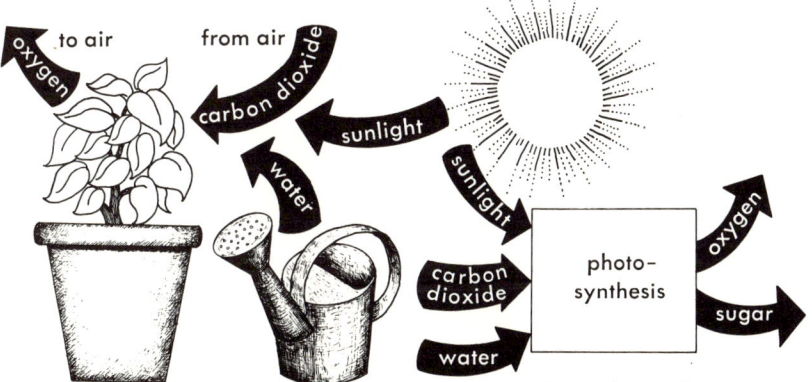

A green plant changes the environment by taking from it sunshine, water and carbon dioxide and giving out oxygen while it makes sugar.

Cycles in Nature

The process of respiration goes on in plants as well as animals. There is also a respiration of soil that takes place as the remains of dead animals and plants in the soil decay. All three kinds of respiration take oxygen out of the air and put carbon dioxide and water into it. In spite of this fact, the air never loses all of its oxygen, because photosynthesis puts back into the air the oxygen that respiration takes out. Respiration doesn't destroy the oxygen it takes out of the air. It stores it as hidden oxygen in molecules of carbon dioxide and water. Photosynthesis doesn't make new oxygen. It frees some of the hidden oxygen from the small molecules that are combined to make sugar. As a result, the oxygen moves in a cycle or circle of changes: Free oxygen is changed to hidden oxygen, and then hidden oxygen is changed back into free oxygen. Because of the working of this oxygen cycle, the amount of oxygen in the air is kept at a fairly steady level of two parts of oxygen in every ten parts of air.

There are plants and animals living in seas, lakes and rivers. They, too, carry on respiration and photosynthesis, using oxygen and carbon dioxide that are dissolved in the water. So there is an oxygen cycle in bodies of water like the oxygen cycle in the air. This

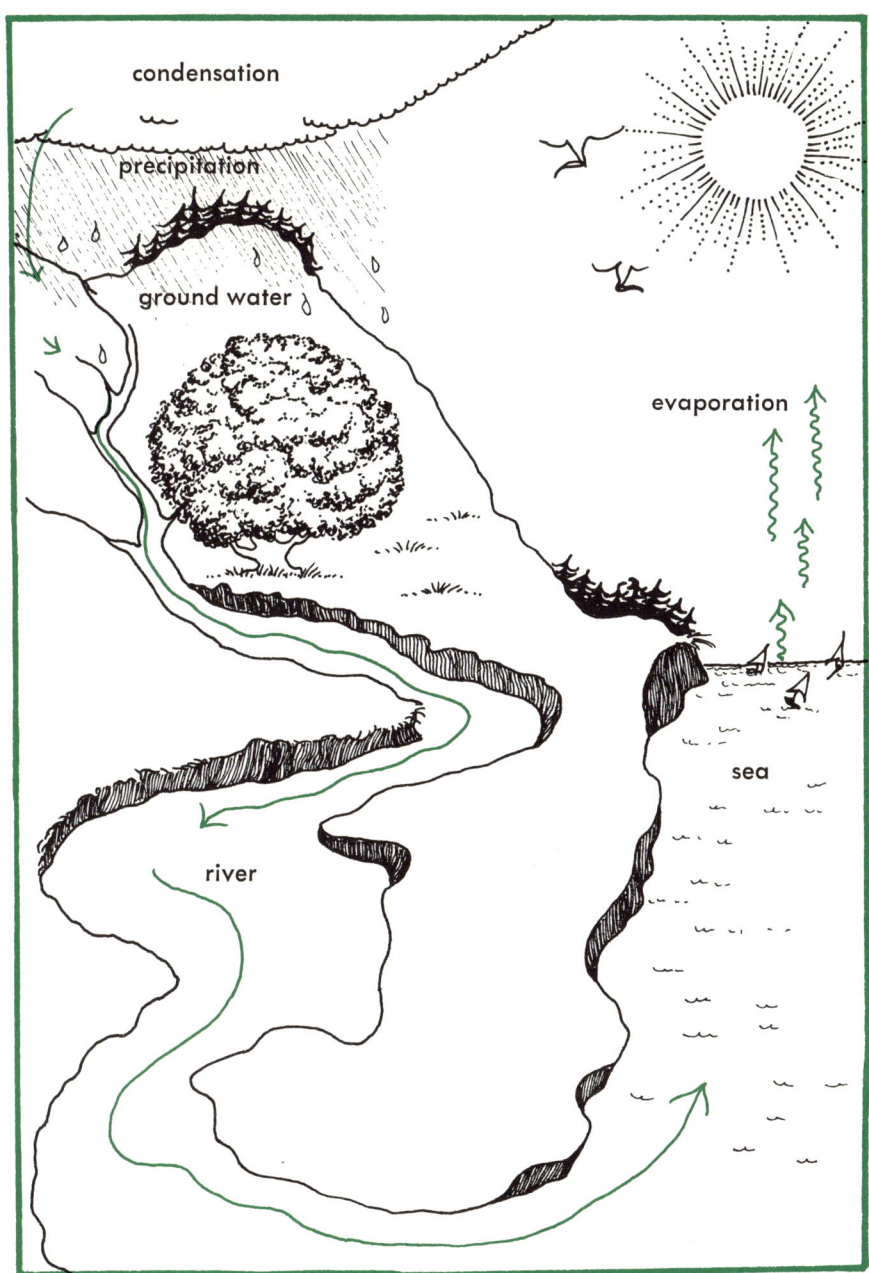

The Water Cycle

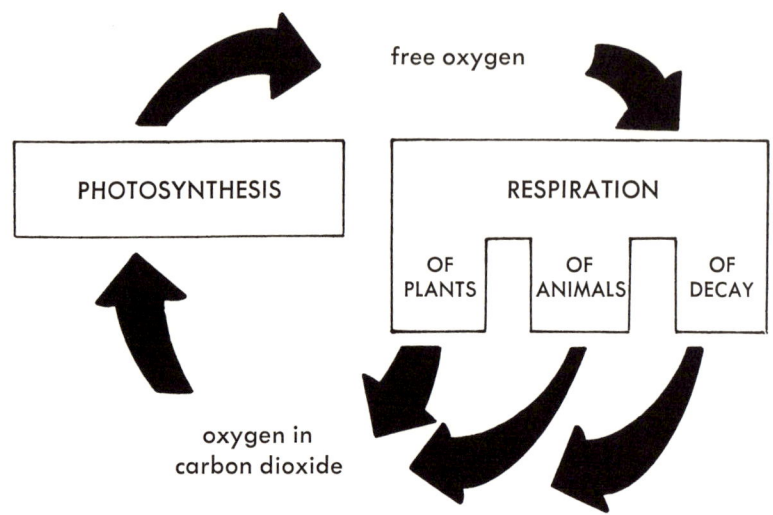

The Oxygen Cycle

cycle can be destroyed if there is too much decaying material in the water, using up the oxygen faster than it can be replaced. What happens then is described on pages 34 and 35.

There are many other cycles in nature besides the oxygen cycle. Among the most important are the carbon cycle, the nitrogen cycle and the water cycle.

In the carbon cycle, carbon hidden in carbon dioxide is stored by photosynthesis in large molecules made by green plants. Then some of it is stored in the bodies of animals that eat the plants. Then some is stored in the soil when the plants or animals die. Then respiration of plants, animals and soil returns the carbon to the state of hidden carbon stored in carbon dioxide.

In the nitrogen cycle, free nitrogen in the air is changed by certain kinds of bacteria into a hidden form, in which it becomes part of the food for plants, which may then become food for animals. When dead plants and animals decay, a complicated series of chemical actions changes the hidden nitrogen into free nitrogen and returns it to the air.

In the water cycle, water rises into the air when it evaporates from seas, lakes, rivers, the ground and living things. Then, after being stored for a while in clouds, it falls to the ground as rain or snow, and returns through streams and rivers to lakes or to the sea.

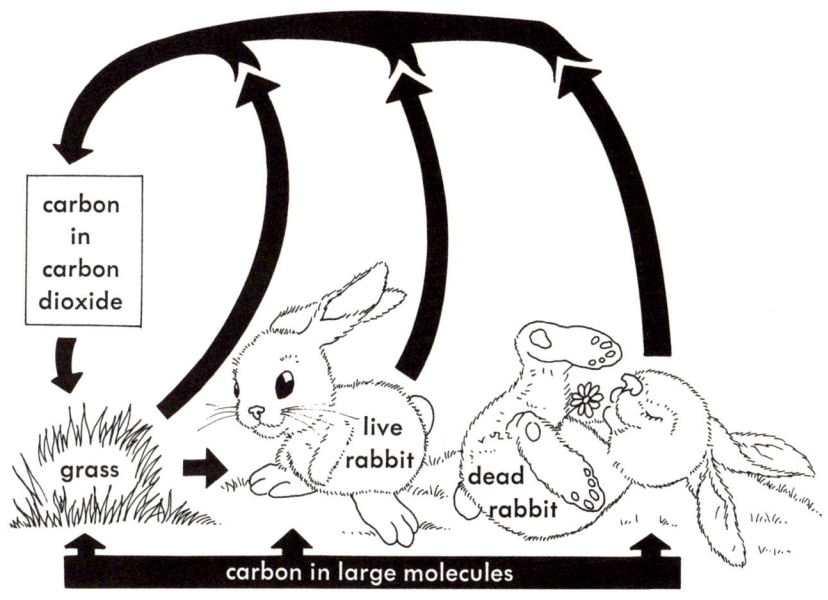

The Carbon Cycle

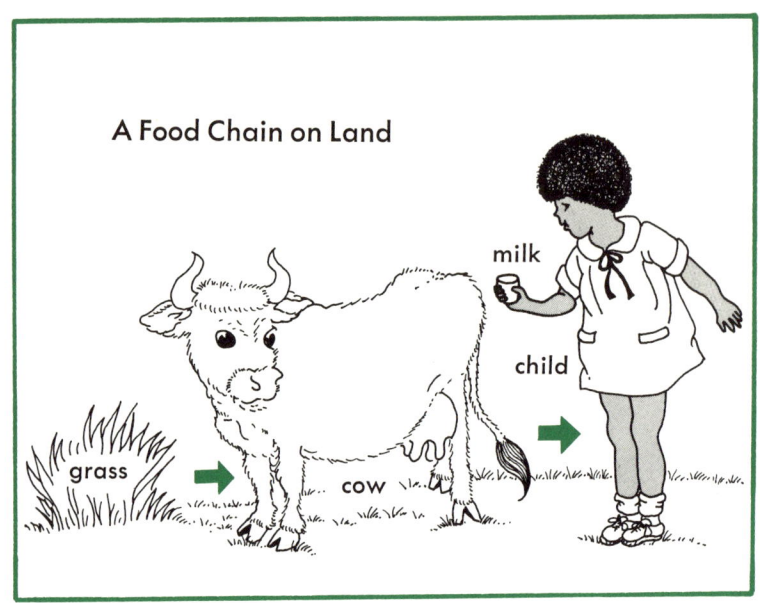

Food Chains

After carbon has been built into the body of a green plant by photosynthesis, it may pass through many living bodies before being returned to the air. When grass is eaten by a cow, the carbon in the grass becomes part of the body of the cow, and some of it is in the milk produced by the cow. If a child drinks the milk, the carbon becomes part of the body of the child. This is an example of a food chain, where each link of the chain is food for the next link. There are two kinds of food chains, *browsing chains* and *decay chains*. The chain, grass–cow–child, is an example of a browsing chain. A decay chain begins when such things as dead animals or plants and body wastes of animals are "eaten" by bacteria.

Both kinds of food chain exist in the sea as well as on land. In the sea, a browsing chain begins with tiny green plants that carry on photosynthesis. These plants are eaten by tiny animals which, in turn, are eaten by small fishes. The small fishes are eaten by larger fishes, and so on.

It isn't only carbon that is passed from one living body to another in a browsing chain. Any chemical that enters the body of a plant enters a browsing chain and is passed from the eaten body to the eater.

A Food Chain in the Sea

Connections and Balance in Nature

Each part of the environment, whether it is living, like a tree, or not living, like the air, is connected with some other parts in many ways. The air is connected to trees, because trees take carbon dioxide out of the air and put oxygen into it. Cows are connected to grass because they eat it. Some things are connected by the cycles described on pages 10 to 13. Some are connected by food chains. Others are connected because they compete with each other for sunlight, air or food.

There are changes going on in nature all the time. Oxygen and carbon dioxide displace each other in the air. Water evaporates from the sea, rains to the ground and flows back to the sea. Plants grow and are eaten. Some animals eat other animals. Because of the connections in nature, each of these changes meshes with others, like the moving gears in a clock. Usually the changes balance each other. While photosynthesis takes carbon dioxide out of the air, respiration puts it back. While one process makes big molecules, another process breaks them up into small molecules again. However, sometimes an unusual change in the environment destroys the balance, and strange things begin to happen. For example, in England an outbreak of disease once killed almost all of the rabbits. The foxes, owls and hawks who used to eat the rabbits had to find something else to eat instead. So they began raiding poultry yards, killing and eating many chickens, ducks and geese.

Man's Work

Man, like any other animal, interferes with the environment by breathing, eating and producing body wastes. But man also interferes with the environment in a way that no other animal does. Man *works*. He doesn't merely use what he finds in nature. He changes what he finds in nature to carry out his own purposes.

For about ten thousand years man has been a farmer, planting some crops for food and some, like flax, for use in making things. To increase the amount of farmland, he has cut down forests, drained swamps and irrigated deserts. He developed crafts for making useful things out of materials taken from nature. He used wood cut from trees to build houses. He used clay to make pots and reeds to make baskets. He learned how to spin and weave wool and other fibers to make cloth.

With a food supply assured by farming, some people began to live in cities. They constructed great buildings and pyramids. They learned how to make metals out of certain kinds of earth. Increased productive activity began to make great changes in the environment. For example, about three thousand years ago, a force of eighty thousand men cut down cedar trees in the hills of Lebanon to make lumber for King Solomon of Israel. Before long these hills were completely bare.

Machines, Power and Chemistry

About two hundred years ago, a great change took place in England in the way people did their work. Instead of spinning thread and weaving cloth by hand, they began to do these jobs by machines invented for the purpose. The power to drive the machines was produced by burning coal. Since then the use of power-driven machinery has spread to other industries and other countries. The invention of the steam engine led to the construction of railroads. The need for iron for the railroads and for machinery led to a rapid growth of the steel industry. The development of practical ways to make and use electricity made it possible to send power through wires over great distances. To produce the electrical power, larger and larger amounts of coal and oil had to be burned. The invention of the automobile, which burns gasoline obtained from oil, led to digging oil wells in many parts of the world and shipping oil over the oceans in great tankers. A chemical industry grew up, in which new chemicals not found in nature were produced. As a result of the rapid growth of industry, the United States alone, in 1971, burned 500 million tons of coal and 5 million barrels of oil. Nearly 2 million different chemicals are now known, and a quarter of a million new ones are made each year, with about five hundred of them being put to use on a large scale in industry and farming.

Population and Its Limits

Every animal except man has its special place in the environment. For example, some animals live in jungles and some in open plains. But man, through his work, has spread to every part of the environment and can live in either the jungle or the plains.

The population of each animal is controlled by two things: the animal's food supply and its natural enemies. If a rabbit population grows faster than its food supply, some of the rabbits starve to death and the population stops growing. Also, as the rabbit population grows, the foxes who eat them are better fed.

Then the fox population grows, and more rabbits are eaten. So, in the long run, each animal population remains about the same.

However, man has been an exception to this rule. Through his work in farming and industry he is better able to feed himself and protect himself than other animals are. As a result, the human population has been growing at a faster and faster rate. The world population of mankind ten thousand years ago was between 5 and 10 million. Two thousand years ago it was about 300 million. By the year 1650 it had grown to 500 million, and it reached 1,000 million by 1850. In 1975, the world population was 4,000 million and, if it continues to grow at the same rate, it will be 6,000 million in the year 2000.

The continued growth of the human population cannot go on forever. The land area of the earth is about 52 million square miles, and this area does not grow. The population in 1975 was about 80 people for every square mile. Improved farming and new ways of making food can feed more than this number, but probably no more than a few hundred per square mile. Estimates of how many people the earth can support range from 10,000 million to 50,000 million. If the lower figure is correct, the growth of the world population will have to stop in the twenty-first century.

Unwanted Side Effects

Man interferes with the environment to get things that he wants. But, since every part of the environment is connected to others, man's interference has side effects on other parts of the environment, sometimes with consequences that man doesn't want.

For example, the people in Africa killed many leopards, partly to protect themselves against attack and partly for sport. The leopards used to kill and eat baboons, and this kept the baboon population small.

An Eroded Hill That Has No Trees

But when increasing numbers of leopards were killed, the baboon population began to grow, and soon the hungry baboons were raiding farms and destroying crops.

Some kinds of man-made damage to the environment can be healed by nature when the damage is small. But when the damage is done on a large scale, nature's healing powers, too, are destroyed and the damage spreads. For example, if a few trees are cut down in a forest, the nearby trees drop seeds into the cleared space, and new trees grow up to fill it. But if a whole forest is cut down on a hillside, the hill loses more than the forest. First, it loses the new soil that might have formed from fallen leaves and branches had the forest remained. Second, it begins to lose water. Forest soil can hold 500 inches of rainfall that soaks into the ground. When the forest is cut down, the soil loses this holding power. Then rainwater runs off the hillside, instead of being held in the ground. The flowing water carries soil with it, so the hillside loses old soil as well as water. Before long, what was a tree-covered hillside has become a naked desert.

With the growth of power-driven industry in the last hundred years, man has been interfering more and more with the environment and has produced more and more unwanted side effects.

Vanishing Species

Just as too much cutting of trees can destroy a forest, too much hunting of animals can destroy a whole species. At one time there were large herds of buffalo in the western plains of the United States. The Indians hunted them for meat, hides and other products. As older buffalo were killed, young calves grew up to take their place, so, from year to year, the herds remained the same size. But, after the white man began to settle on the plains and the railroads were built, a large-scale slaughter of buffalo took place. The mass killing of buffalo went on partly to clear the plains for farmland, partly to starve the Indians in the white man's war

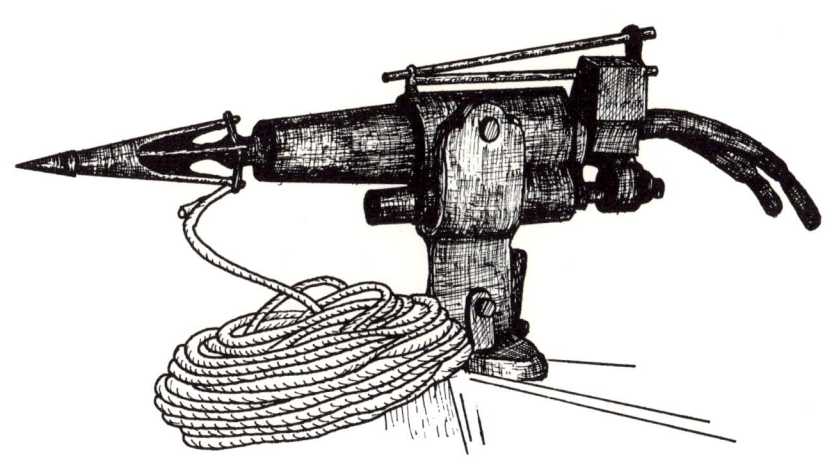

Whaling Harpoon

against them, and partly to sell buffalo meat and hide at a profit. Before 1840, there were 50 million buffaloes on the plains. By 1890, there were almost none. Now buffaloes no longer exist in the United States as sources of food and hides. They exist only as curiosities in zoos and national parks.

Whales are also hunted for meat and other products. In Japan, one out of every ten pounds of meat eaten is whale meat. Too much hunting threatens to wipe out the whale as it wiped out the buffalo. The whale can still be saved if all countries agree to limit the number of whales they kill each year. Then it will still be possible to hunt for whales in the future to get whale meat, whale oil and other useful things.

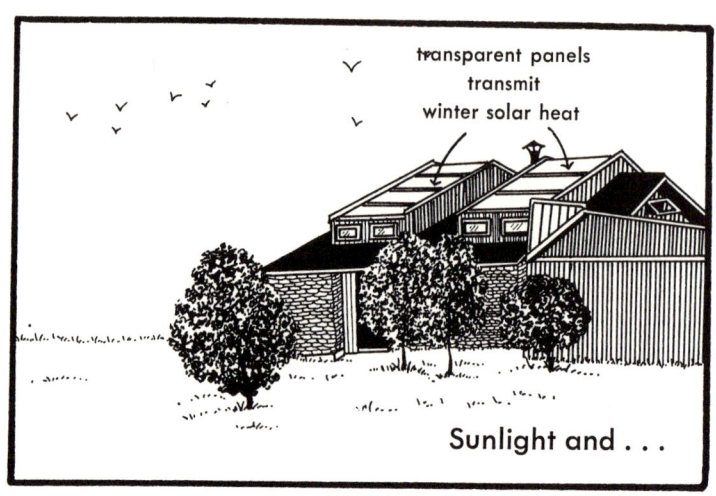

Used-up Treasures

Some of the things that we take from nature's storehouse of treasures are *renewable:* They are or can be replaced as fast as they are used up. Sunshine is renewable. A fresh supply comes in from the sun every day. Oxygen is renewable, as long as there are enough trees to make it as fast as living things take it out of the air. Trees are renewable if we plant a tree every time we cut one down.

But some of the things we take from nature are not renewable. Petroleum oil is not renewable. We are burning it so fast in cars, trucks, planes and furnaces that the world's supply will be used up in from thirty to sixty years. Nature cannot replace it, because the natural process that makes oil takes hundreds of millions of years. For this reason, it is necessary to save oil by not using it wastefully, and to find substitutes for oil. A kind of oil can be made from coal. Coal, too, is not renewable, but at least there is enough coal in the ground to last hundreds of years. For some purposes, such as heating homes or pumping water, oil can be replaced by sunlight and wind power, both of which are renewable.

Metal ores, the earths from which metals are made, are not renewable. We can avoid using up the world's supply of metal ores by imitating nature. Just as nature recovers oxygen from the used oxygen in carbon dioxide, we can recover a metal from the used objects that contain it. If scrap metal is collected and melted down in a furnace, a fresh supply of usable metal can be made from it. This process is called *recycling*.

... Wind Are Renewable Power Supplies

Hogging the World's Resources

Most of the world's industries that use power-driven machinery are in the United States, Japan, the Soviet Union, England, Germany and other countries of Europe. These are known as the *industrialized* countries, or the *developed* countries. The use of power-driven machinery makes it possible for the industrially developed countries to produce large quantities of goods for the use of their people. To produce these goods they use large quantities of fuel and minerals. The developed countries have only one-fourth of the world's population, but they consume more than nine-tenths of the world's minerals and two-thirds of the world's oil. The United States alone, which has only one-eighteenth of the world's population, uses one-fourth of the oil. On the average, each person in a developed country consumes one hundred times as much as a person in an underdeveloped country, using up in six months the amount of energy and raw materials that a person in an underdeveloped country uses in a lifetime.

Since the end of World War II, many countries of Africa and Asia, which used to be controlled by other countries, have become independent. Now they, and other underdeveloped countries, have begun to de-

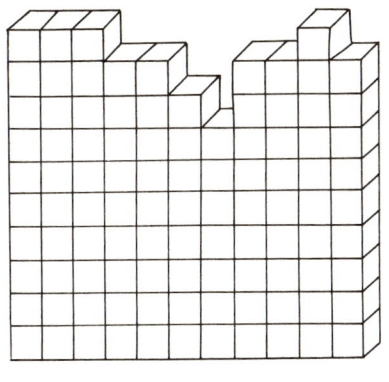

share of world population of developed countries

share of world's energy and raw materials consumed by developed countries

share of world population of undeveloped countries

share of world's energy and raw materials consumed by undeveloped countries

velop their industries so that they, too, may produce more goods for their people. As *developing* countries, they will increase from year to year the amount of fuel and minerals that they will use. With the developed countries using most of them, and the developing countries using more and more, a shortage of fuels and of some minerals is bound to develop. It will be necessary for the countries of the world to make agreements on how much of the world's resources each country may use, to be sure that each country gets a fair share and that enough is left for future generations.

Fouling the Air

Factories and internal-combustion engines have helped make the developed countries rich. The factories produce large quantities of goods, and internal-combustion engines in cars, trucks, locomotives and planes carry the goods and people quickly from one place to another. At the same time, however, factories and internal-combustion engines produce the unwanted side effect of dirtying the air. Factory chimneys pour out smoke, including the gas sulfur dioxide. Cars and trucks, through their exhaust pipes, pour out several kinds of poisonous gases, including carbon monoxide, hydrocarbons and nitrous oxides. When only small amounts of these gases were put into the air, they did little harm. But now the amounts are large enough to cause property damage, illness and even death.

In 1956 the factories and cars and trucks in the United States poured out 42 million tons of *pollutants,* or things that dirty the air. Half of them came from cars, trucks and planes. In any big city, the daily output of pollutants by a thousand cars is over three tons of carbon monoxide and from a quarter of a ton to half a ton of other harmful gases. These gases become especially dangerous when there is an *inversion layer* of air overhead, a layer of warm air over the cooler air

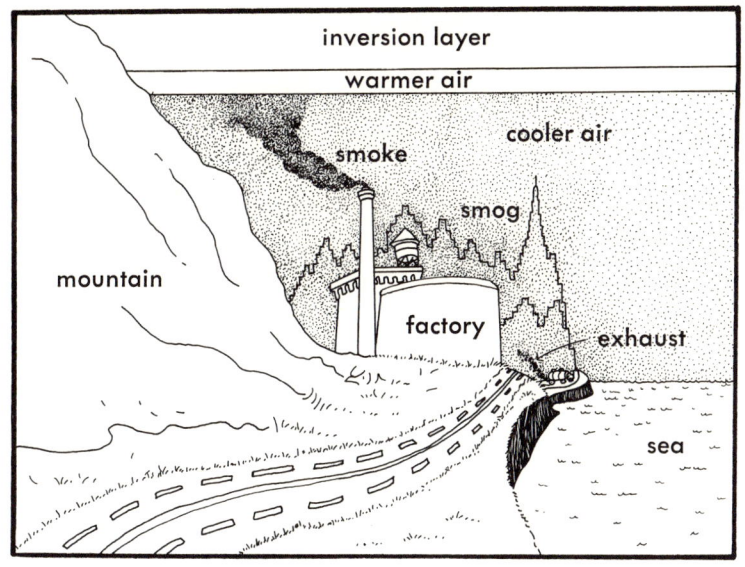

at the ground. The inversion layer is like a lid on the air below it. It traps the harmful gases below, preventing them from rising. The hazy air that results is called *smog*. A smog in London, in 1952, caused over four thousand deaths. A New York smog, in 1966, caused 168 deaths.

In order to begin to clean up the air, some cities forbid the release of smoke that contains sulfur dioxide. In 1963, the United States Congress passed a Clean Air Act requiring automobile manufacturers to put on each new car a device that removes some of the pollutants from the automobile exhaust. However, with more and more cars being used, by 1980 the air may be as dirty again as it was before.

Spoiled Water

Farms, industries and cities are all useful, but they have side effects that help to spoil the water in lakes and streams.

The crops a farmer grows take plant food out of the soil. To be able to grow the next crop, the farmer puts fertilizer into the soil. Some of this fertilizer is washed out of the soil by rainwater, which then carries it into nearby lakes and streams. When there is too much fertilizer in a lake, the tiny floating green plants known as *algae* grow rapidly to form a thick layer on the surface of the lake. The thick layer cuts off the sunlight from the algae beneath the surface. They die and begin to decay. The process of decay uses up the oxygen in the water. Then the fish in the water die, all plants in the water die and even the bacteria of decay die. This has happened to some lakes and streams in many parts of the world.

In some cities, sewers pour the city wastes into nearby streams and lakes. Sewer wastes contain fertilizer, including phosphates used in some laundry detergents. These, added to the fertilizer washed out of farmland, may destroy the life in the lakes and streams.

Some industrial wastes, too, are dumped into lakes and streams. Sometimes, while a factory pours wastes, including some poisons, into a river upstream, a city

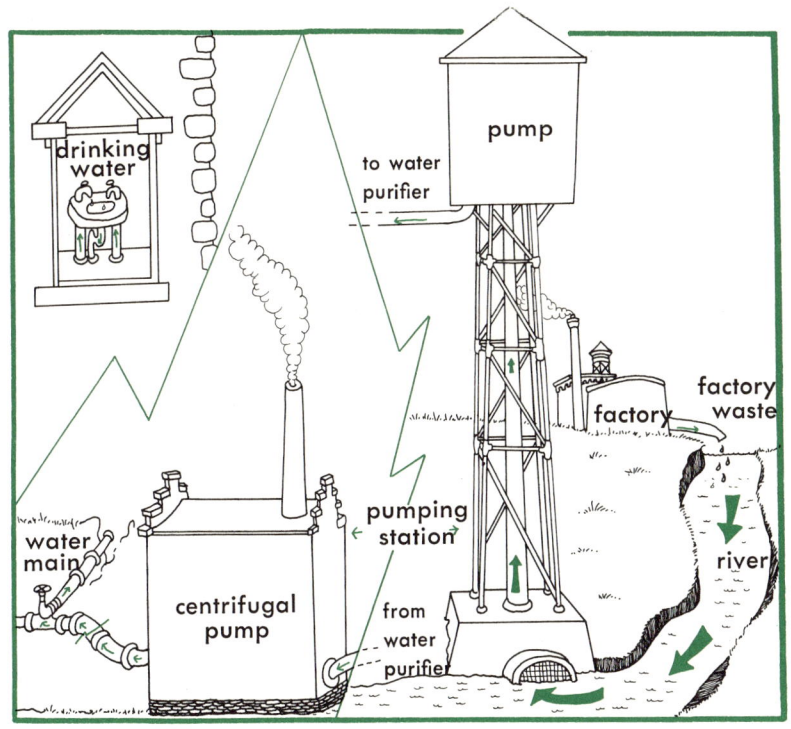

How Poisons in Factory Wastes Can Get into Drinking Water

downstream pumps water out of the river for drinking. This is happening, for example, in the Mississippi River, where mercury from factory wastes has found its way into drinking water.

Water pollution in the United States in 1964 killed over 18 million fish, and also made many people ill.

In 1965, the United States government passed a Water Quality Act, taking the first steps toward finding ways of keeping wastes and fertilizers out of the water.

Destruction of Soil

Nature makes soil out of rock. Heat, cold, wind and water make the face of a rock crack and crumble to form earth. Then the growth of plants in this earth and the decay of dead plants and animals add carbon and nitrogen compounds to it to make it a rich soil. It may take a long time to make a rich soil, but bad use of the soil can destroy it quickly.

In the mountains of Central America, much good soil was destroyed by cut-and-burn farming. The Indians cleared forest land for farming by cutting and burning trees. Cutting down too many trees led to loss of soil, as explained on page 25.

In places where sheep and goats and cattle are raised, soil is sometimes destroyed by too much grazing of the grass. Overgrazing of grass kills the grass. Then the grass is unable to hold the soil in place, and wind and rain carry the soil away.

In the dry Southwest of the United States, some soil was destroyed when grassland that is good for grazing was plowed up to plant crops. Windstorms after a long dry spell blew the topsoil away.

Bad plowing can destroy soil. If a field on a hill is plowed into furrows running up and down the slope, rainwater flows downhill quickly in the furrows and carries soil away.

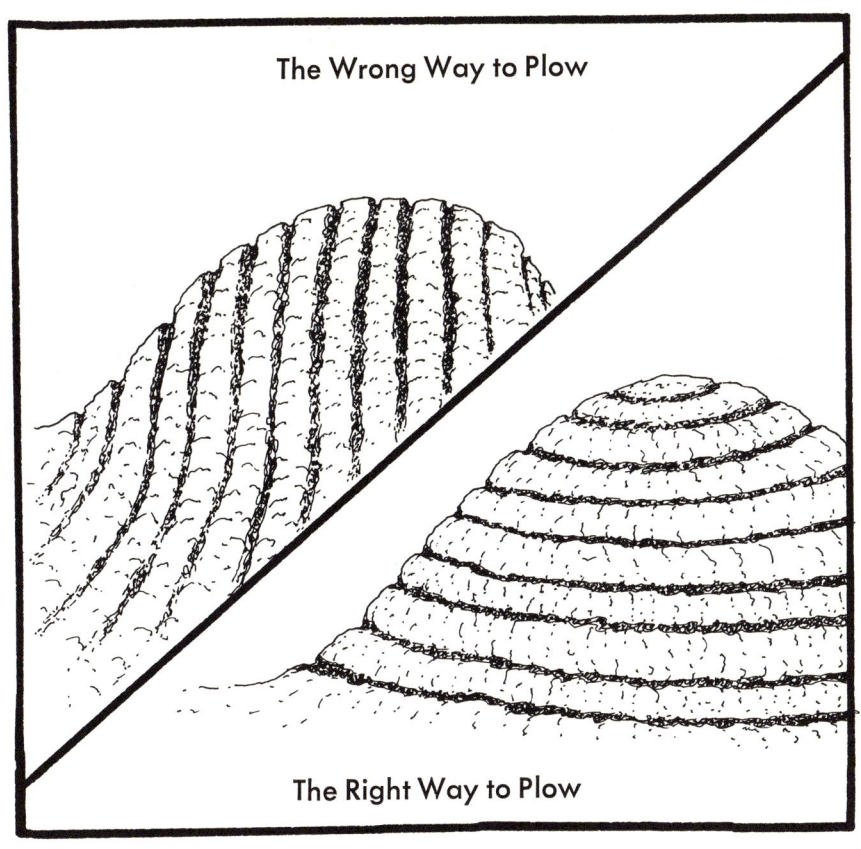

Strip-mining for coal and ores destroys soil.

The growth of cities and suburbs also destroys soil by covering it with buildings and pavements. We cannot prevent the loss of soil to cities, but we can prevent and make up for the other losses. Soil can be saved by planting trees on bare hills, putting soil back after strip-mining, stopping the plowing of dry grasslands, avoiding overgrazing and plowing only level furrows on hills.

Mountains of Rubbish

An industrial country like the United States produces large quantities of things made of steel, including automobiles, dishwashers and stoves. When these are worn out, they are thrown away.

Tons of paper are used to print newspapers and magazines. After these are read, they are thrown away.

Food, beer and soft drinks are packed and sold in millions of metal cans and bottles. After the contents are used, the cans and bottles are thrown away.

Each city throws away in a year enough junked cars, paper, bottles and cans to make a small mountain of rubbish. To avoid being buried under its rubbish, most communities bury the rubbish instead. But this takes time, effort and money.

A better way to deal with rubbish is to recycle it. Steel can be melted again and reused. Bottles can be crushed and melted to make new bottles. Old paper can be used to make new paper. It is even possible to find other uses for rubbish. Crushed glass can be made into building blocks. Excess paper can be turned into sugar and fuel.

In some small communities, volunteers have set up recycling centers, where people take their newspapers, bottles and tin cans. In large cities, the recycling collections can be organized by the departments responsible for collecting garbage.

Two states, Vermont and Oregon, have banned the use of nonreturnable bottles and cans for beer and soft drinks. Only deposit bottles are permitted. Since most people return their bottles and cans to get their deposit money back, not many bottles and cans are thrown away in these states at parks, beaches and roadsides.

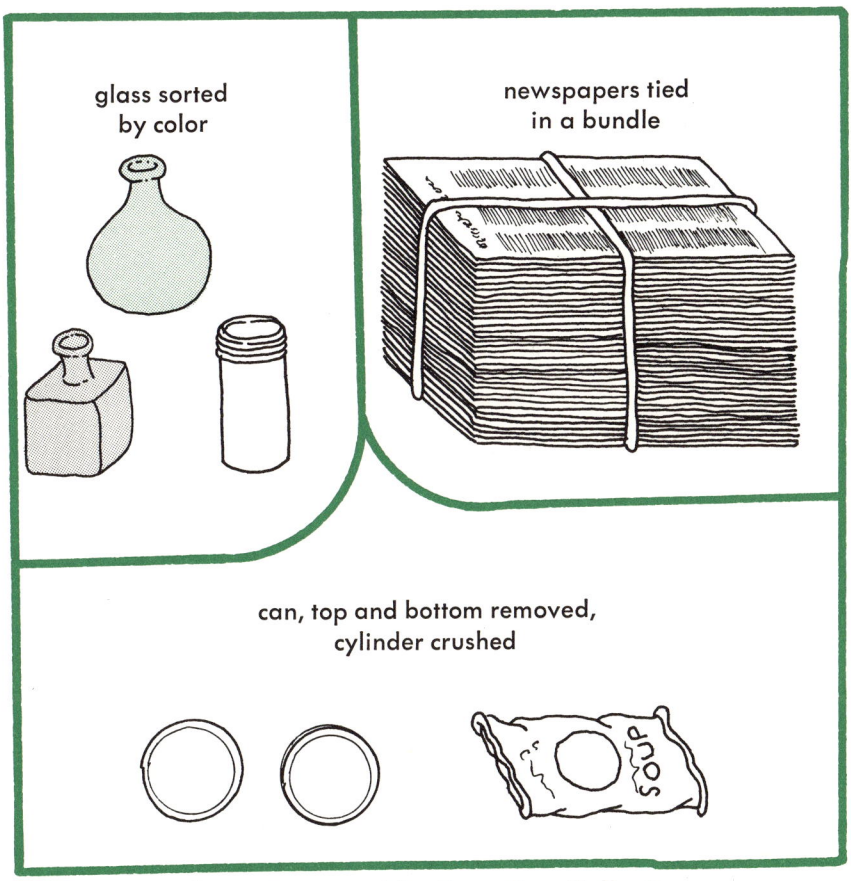

Things Prepared for Recycling Collection

Man-made Poisons

Chemists have constructed thousands of new compounds not found in nature. Uses have been found for hundreds of them in factories, homes and farms. Some of these useful chemicals have turned out to have harmful side effects, so their use had to be stopped. The best-known example of a useful chemical with harmful side effects is DDT.

DDT was invented in 1874. In 1939, it was found that DDT can kill some insects. During World War II, it was used successfully to kill body lice, which are carriers of the disease, *typhus*. In 1944, the use of DDT quickly ended an outbreak of typhus in Naples and saved many lives. After the war, DDT was used throughout the world to kill mosquitoes which carry the disease, *malaria*. Since then it has prevented 500 million deaths due to malaria. In the United States, DDT was also used widely in homes and barns to kill flies and on farms to kill insects that destroy crops. But DDT also had other effects that were not good.

First, DDT didn't kill only harmful insects. It killed useful insects, too, such as bees and ladybugs.

Second, DDT didn't kill only insects. It also began to kill birds, fish, frogs, snakes, crayfish and other animals. It got to these animals through the food chains in nature. Birds and frogs eating poisoned insects were poisoned, too. The poison was even entering the bodies of people. Cows in barns that were sprayed with DDT stored DDT in their fat. Some of it, in their milk, was passed on to people who drank the milk.

Because DDT could kill many animals and could make people ill, the United States government banned the general use of DDT in 1972.

Other Environmental Problems

To show the great variety of ways in which man's work may produce unwanted changes, we describe three more interesting examples.

In the late 1940s, detergents were widely used for the first time in place of soap. In a few years, some lakes and streams and even wells of drinking water were filling up with suds. This happened because the large molecules of these detergents did not decay in the ground. The solution to this problem was to change

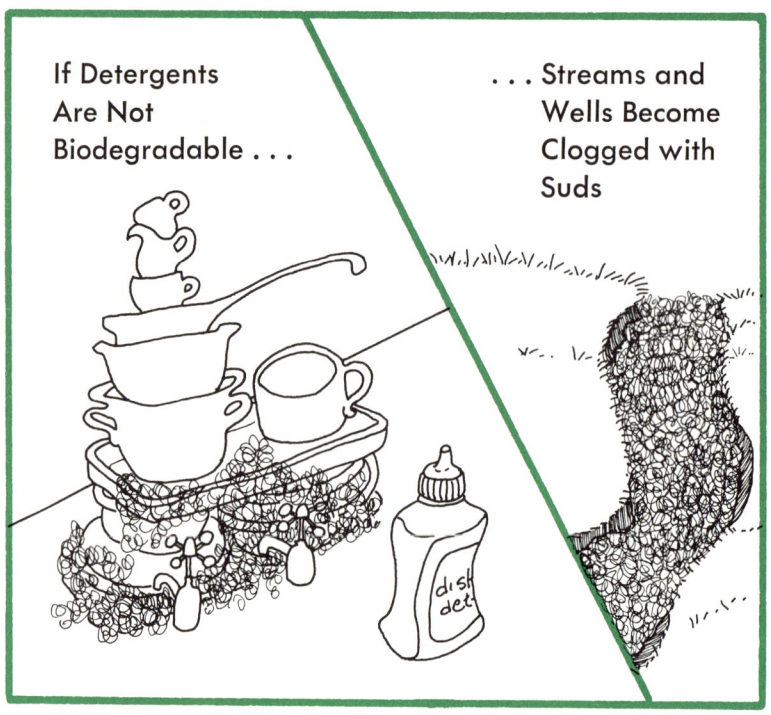

If Detergents Are Not Biodegradable . . .

. . . Streams and Wells Become Clogged with Suds

the chemical formulas of these detergents to make them *biodegradable,* that is, able to be broken down by the action of bacteria in the ground.

For about one hundred years, the developed countries have been burning large quantities of coal and oil. This has increased by one-tenth the amount of carbon dioxide in the air. The carbon dioxide in the air is like a blanket that slows down the loss of heat by the earth. Increasing the carbon dioxide is like thickening the blanket. The result is that the average surface temperature of the earth has been increasing. By the year 2000, it will be almost one degree Fahrenheit higher than it was in 1970. No one knows whether this increase will have any harmful effects.

Some spray cans contain gases called *fluorocarbons.* When these gases are released, they rise in the air to the level of the *ozone* layer. Ozone is a special kind of oxygen that serves as a filter in the air, removing from sunlight most of the ultraviolet rays that are in it. Fluorocarbons in the air may destroy some of the ozone. If a lot of ozone is destroyed, more ultraviolet rays will reach the ground. This is harmful because too much exposure to ultraviolet rays can produce skin cancer. If it is proved that fluorocarbons do indeed destroy part of the ozone layer, their use in spray cans will be banned.

Protecting the Environment

Since the work we do sometimes tends to damage the environment, it is necessary to plan our work so that we damage the environment less, and heal it where it has already been damaged. Experience during the last hundred years has taught us some important rules for protecting the environment.

About things we take from the environment: If the supply is limited, don't use it wastefully. Leave some for use later and also for the use of other countries and of future generations. If possible, switch to substitutes that have renewable supplies. For example, for energy, use less oil, and learn how to use more of the renewable supplies of sunlight, wind and wood.

About things we put into the environment: Stop putting poisons into the air, water and soil. Remove them from factory smoke and automobile exhausts. Test new chemicals, before they are used, to be sure they do not have harmful side effects.

Use nature's methods: Recycle used materials such as metal, bottles and paper. This makes the supply renewable and solves the problem of disposing of solid wastes. Plant trees to replace those that are cut down. As far as possible, make new chemicals biodegradable. To kill insect and other animal pests, use chemicals less and their natural enemies more.

Government action: Laws are needed to protect the air, water and soil, and these laws should be enforced. Raw sewage should not be dumped into lakes and streams.

International cooperation: International agreements are needed to plan the use of resources that are not renewable and to clean up the air and water that cross national boundaries.

What Everyone Can Do

Everybody can help to protect the environment. You can help by writing to your government representatives asking them to pass the laws that are needed. You can also help by doing some things on your own and in cooperation with your neighbors.

To save fuel: Turn off the lights when you leave a room. Don't overheat rooms in the wintertime.

To reduce air pollution and save gasoline: For short trips, walk or use a bicycle instead of riding by car. For longer trips, share car rides with your neighbors, so that one car is used instead of two or three.

To reduce water pollution: Use only detergents that are biodegradable and have no phosphates.

To keep public places clean: Don't litter with paper, cans or bottles. Use waste receptacles.

To reduce solid waste: Join with your neighbors to organize a recycling center where paper, cans and bottles may be brought to be picked up for reuse.

If governments pass and enforce the laws that are needed, and each person helps in every way possible, we can restore a clean environment in which to live, work and play.

Word List

Biodegradable—Able to be broken down by soil bacteria into small molecules that can fit into nature's cycles.

Cycle—A series of changes that return to the condition they started from, so that the changes are repeated over and over again.

Developed country—A country with a high level of industrial production using power-driven machinery.

Inversion layer—A layer of warm air over cooler air at the ground.

Photosynthesis—The process in green plants that uses energy from sunlight to combine carbon dioxide and water to make sugar.

Pollutants—Things that dirty the air, water or ground.

Recycling—Collecting and reworking used things so that the materials out of which they are made may be used again.

Respiration—The process that goes on in most living things to produce the energy of life by "burning" in oxygen large molecules that contain carbon.

Smog—A hazy condition of the air when it contains many large molecules from smoke and automobile exhaust.

Strip-mining—Mining done by stripping off layer after layer of the ground, starting at the surface.

Index

Air, 6, 10, 13–14, 16, 32, 44–45
Automobile, 20, 44

Bacteria, 13, 15, 34
Balance, 16–17, 22–23
Biodegradable, 43–44, 46–47
Bottles, 38–39, 46
Browsing chain, 14

Cans, 38, 46
Carbon, 8, 12–13, 36
Carbon dioxide, 9–10, 16, 43
Carbon monoxide, 32
Chemistry, 20, 44
Cities, 18, 37
Coal, 7, 20, 29, 37, 43
Connections in nature, 16–17
Cycles, 10–13, 16, 47

Damage, man-made, 25, 36, 40–44
DDT, 40–41
Decay, 12, 14, 34, 36
Detergents, 34, 42–43, 46
Developed countries, 30, 43, 47
Developing countries, 31

Energy, 8–9, 44

Factories, 32
Farming, 18, 21, 23, 36
Fertilizer, 34–35
Fluorocarbons, 43
Food, 6, 16, 18, 22–23, 34
Food chain, 14–15, 40–41
Fuel, 30–31, 38, 46

Gasoline, 20, 46

Industry, 21, 23, 25
Insecticides, 40–41
Insects, 40, 44
Interference, 8–9, 18, 24
Inversion layer, 32, 47

Machines, 20–21, 30

Metals, 7, 18, 29
Minerals, 30–31

Natural enemies, 22, 44
Nitrogen, 6, 13, 36
Nitrous oxides, 32

Oil, 7, 20, 29–30, 43–44
Ores, 7, 29, 37
Overgrazing, 36–37
Oxygen, 6, 9, 10, 12, 16
Ozone layer, 43

Phosphates, 34, 46
Photosynthesis, 9–12, 14–16, 47
Plants, 6, 9–10, 12, 14, 16
Plowing, 36–37
Poisons, 40–41, 44
Pollution, 32–35, 46–47
Population, 22–23, 30
Power, 7, 20, 25, 29–30
Protection of environment, 44–47

Recycling, 29, 38, 44, 46–47
Renewable resources, 28–29, 44–45
Respiration, 8–10, 12, 16, 47

Side effects, 23, 32, 34, 44
Smog, 33, 47
Smoke, 32–33, 44
Soil, 7, 10, 12, 25, 34, 36–37, 44–45
Spray cans, 43
Strip-mining, 37, 47
Sugar, 9–10, 38
Sulfur dioxide, 32, 33
Sunlight, 9, 16, 28, 34

Temperature of the earth, 43
Trees, 16, 18, 25, 28, 36–37, 44

Wastes, 34–35, 38–39, 45–46
Water, 6, 9, 10, 11, 13, 16, 25, 34–35, 44–45
Work, 18–21, 23, 25

10873

574.5 Adler, Irving
ADL   The environment

DATE			

HOOSIC VALLEY ELEMENTARY SCHOOL
SCHAGHTICOKE, NEW YORK

© THE BAKER & TAYLOR CO.